WINERY

Four Seasons in the Vineyard

By Richard A. Occhetti

ISBN 978-1-62806-164-2   (print | paperback)

Library of Congress Control Number 2018940125

Published by Salt Water Media
29 Broad Street, Suite 104
Berlin, MD 21811
www.saltwatermedia.com

Salt-Water
MEDIA

Cover images and interior photographs by the author,
with the exception of the photo of Peggy Raley-Ward on page 10,
which is provided courtesy of Kevin Fleming.

Dedication

To Christian, George, Jessica, Mignon, Nelson, Paula, Shane, Tony, and Wally, my fellow workers in the vineyard; and to Suzette, who always sees to it that I have enough wine.

Acknowlegdments

Some forty years ago, my good friend and neighbor Don Cullingham introduced me to wine-making. Knowing little more about the subject than I did (which is to say, practically nothing) Don bought himself a top-of-the line wine press from a company in Italy, and began planting enough grapevines to fill most of his backyard. I was fascinated, and immediately followed suit. The two of us learned by doing, and by visiting wineries whenever either of us happened to be anyplace in the world where wine was made. Over time we progressed to the point of realizing how much we didn't know, and turned to the excellent books of Phillip M. Wagner to explain the mysteries.

Thus, I learned enough to give myself many years of satisfying diversion and to make some passable wine. But this is quite a different thing from making wine good enough to sell... and doing it on a commercial scale. I could never have written about that without the things I have learned from my new friends, Peggy Raley-Ward and Mike Reese. Peggy is among the leading authorities on wine-making and viticulture in America today. She generously shared her time and expertise with me during the writing of this book. Mike is an innovative hands-on craftsman, who has translated the skills needed for producing a masterpiece into a system for turning out masterpieces in quantity. Watching him made this an easy book to write.

R. A. Occhetti
Lewes, Delaware
December 2017

THE PLACE and THE PEOPLE

Just outside the charming little seacoast town of Lewes, Delaware, there is a winery – Nassau Valley Vineyards, by name. It sits on some twenty acres of prime real estate convenient to shopping and dining and a scant four miles from the beach. By any economic logic, this ought to be a development of luxurious leisure homes. Instead, it is an oasis of sunlight and greenery, where rustic buildings punctuate orderly rows of grapevines in a gently rolling landscape. This book is about how this winery came to be and what it is like to spend a year in this improbable place practicing an art nearly as old as civilization itself.

Early in the 1950's an instinctive entrepenour named Robert A. Raley spotted some of the business opportunities awaiting discovery in "lower, slower" Delaware. He moved to Lewes from his native Washington, D.C., married a girl whose family roots ran deep in the rural soil, and set up shop as a florist and landscaper. He prospered, and by 1970 was able to buy the big farm which now contains the winery. He leased most of its acreage to other farmers who grew the regional cash crops of corn and soybeans, while he reserved a section to his own use and worked it with his family and a few employees. He planted trees and shrubs for his landscaping business, and grew alfalfa hay for another of his businesses, a herd of beef cattle, which he kept on land he owned nearby. A few of the landscaping products can still be seen at the winery in mature groves of neatly-spaced trees; the cattle were sold when Bob died in 2013.

The Raley children included a daughter named Peggy, who was not particularly fond of working on a farm. She made her escape by going away to college.

To help with tuition, she took a job as a photographer for the periodical published by "*Les Amis du Vin*," an organization devoted to advancing the public's awareness and appreciation of fine wines. Peggy had no particular interest in wine when she started, and was too young to drink legally anyway, but the job quickly advanced her awareness and appreciation. In short order, she went from taking pictures to writing articles to living in the storied French wine region of Bordeaux, all the while learning about wine from some of the world's best wine-makers. Inevitably, the day came when she decided that she would rather make wine of her own than write about other people's. Her father recognized the impulse, and saw in it the way to lure his youngest daughter back home: he would open a winery for her. His plan had only one flaw – commercial wineries could not operate legally in Delaware at the time. Peggy returned to Delaware in 1989 and embarked on a one woman campaign to change the law.

She had already decided that when and if the law let her have her winery, she would grow only classic European wine grapes. This was a choice which most of her friends considered foolhardy, since attempts to grow these grapes in the eastern United States had a long history of failure dating back to colonial times. The poster boy for that long history is none other than Thomas Jefferson, who couldn't keep his vines healthy, even with a team of Italian experts who came from Europe specifically to help him. The conventional wisdom ever since had been that overall conditions in America (with the golden exception of California) just weren't suitable for growing European wine grapes. But Peggy thought differently. She believed that if the specific components of those overall conditions – things like soil composition, insects, and humidity – were dealt with individually, agricultural advancements over the past 200 years might

Peggy Raley-Ward, vintner, jazz singer, and reluctant farm girl
Portrait by Kevin Fleming

enable success. She was prepared to test that belief.

In 1991 the state legislature adopted a bill permitting commercial wineries in Delaware for the first time in the state's history. Peggy had personally drafted the bill, and spent two years working to enact it. Now work could begin in earnest on converting a portion of the farm into a proper winery. A cement floor was poured in the farm's large equipment barn, which became the wine cellar. Small buildings and walkways were erected – some made of wood salvaged from the Rehoboth Beach boardwalk, which had washed away in the nor'easter of 1992. Storage tanks and other wine-making equipment were acquired, grapevines were purchased and planted and a few miles of trellises erected. The work was extensive and the investment staggering, even for a family which already owned the land, and it would be seven to ten years before any income could be expected from the enterprise: four years for the vines to mature sufficiently to produce decent grapes, two more for the wine made from those grapes to age, and who knows how long for people to start buying it. Refusing to become discouraged, Peggy sang professionally in the evenings to earn a living and worked by day in the vineyard, slowly but surely making real progress.

Eventually, the winery grew to the point where Peggy could no longer do all the work herself, so she started hiring wine-makers. When she advertized for a new one in 2007, a muscular twenty-six year old applied for the position. He had been laid off from his construction job when the housing market collapsed in the recession which began that year. Peggy describes hiring Mike Reese as a "leap of faith." He had no professional experience in the wine business, and looked not at all like the middle-aged aristocrats who ran the vineyards of France. But he seemed to possess the intrinsic qualities needed to do the job right. Peggy saw him as a disciplined and determined man with an intense will to succeed, – in her words, plenty of "heart." So she took the leap. She linked him with a reliable mentor, an experienced vintner from an established winery in Virginia, while Mike recalled the lessons he had learned as a child watching his father and grandfather make home-made wine. He also studied everything he could lay his hands on about growing grapes and the science of wine-making. Neither of them has looked back since.

Mike Reese, winemaker

WINTER IN THE VINEYARD

Driving by the vineyard on a January morning, you can see the dormant grapevines -- long rows of plants on a snow-covered field, their spindly branch-es clinging with dry tendrils to the wires of their trellises. A few months earlier these vines produced the vintage that is aging this winter in the winery's cellar; now they rattle around in the chilly breeze, taking the nap which nature affords them before they do it all over again in the next growing season.

Looking through the leafless branches, we have an unobstructed view of nearly the entire vineyard and can clearly see that in shape, size and spacing every vine on the land is essentially identical to every other. This is deliberate, of course, the result of choices Peggy made when the vineyard was planted. Each new vine was trained in such a way as to concentrate all of its energy into grow-ing two upright branches or "canes." When they grew to the height of the lower wire on their trellis, the canes were bent at right angles, in opposite directions to each other, and tied to the wire, thereby producing a shape resembling a capital "T." Vines trained in this manner are generally described as "*cordons*," the French word for ribbon. The principal advantage in training a vine to two cordons is that, should anything happen to one of them, a cane from the other can usually be realigned to take its place in the row.

A cordon may be thought of as the permanent trunk of the grapevine; the dozens of smaller branches growing out of the cordons are the ones which bore the grapes of the previous vintage. All of them must be either partially or entirely removed before winter ends.

The Trellises

Nature intended grapevines to be exuberant growers. They produce canes of extraordinary length, which fan out in every direction, sometimes sprawling on the ground, and just as often climbing into the highest branches of any nearby tree. Obviously something is needed to tame this wild habit of growth and to keep the fruit in a position where it is easy to find and to reach. Hence, the trellis.

There are several types of trellises in use throughout the grape-growing world. The one used here has as its main element a horizontal wire positioned two to three feet off the ground and stretched taut between two vertical posts. The cordons are attached to that wire with plastic ties. As the trellises are hundreds of feet long, there are intermediate posts between the two main ones to keep the wire from sagging. About six feet off the ground, there is a second wire running parallel to the one just described. It is there to give the branches which

bear the grapes something to grab for support when they reach their full length. Between these two, there is another wire which wraps around the entire trellis like a belt so that branches can be tucked into it and held upright as they grow. Unlike the other two, it can be moved up and down to keep pace with the growth of the branches over the summer.

Pruning

Grapevines can live well past fifty years. Each spring, new branches emerge from buds on the branches which were new the spring before, and they in turn produce the buds which will sprout new branches the following year. Fruit forms only on the vine's newest branches, so if this habit of growth is not interrupted, the branches which bear the grapes will be located progressively farther from the plant's roots every year, and the vine will become a tangled thicket with clusters of undernourished grapes scattered all about its distant edges. What the vintner wants, of course, is just the opposite: an orderly vine in which the plant's energy goes into producing plump and healthy grapes positioned where they are easy to care for and to pick. That is why branches must be shortened or "pruned" each winter.

Pruning is labor-intensive and time consuming, but it is not especially complicated on a mature vine. All of the smaller branches on the vine – everything except the cordons themselves – are new growth from the previous summer. The pruner selects those branches which are in places where he wants grapes to grow, and cuts them down to a length of about two inches; then he completely removes all the others. The two-inch pieces of branch which he leaves on the cordons are called "spurs," and each of them typically has two or three buds growing on it. New branches will emerge from those buds when the vine breaks dormancy in the spring, and shortly afterward grapes will grow on them.

The actual pruning at Nassau Valley is a pretty substantial operation because there are more than three thousand grapevines in the vineyard and only a short window of time in which to do the work. It must be done while the vines are dormant, but it is not wise to start cutting too early. Even in the mild climate of coastal Delaware there is the chance that a hard, bud-killing freeze may occur at any time during the winter. Such a freeze seldom kills all the buds on a

vine, but since pruning leaves only a few, it is prudent to let a good part of the winter go by before starting.

The first step in pruning is rough trimming with a gasoline powered hedge clipper. This does not involve any picking and choosing among the branches to determine which spurs to keep and which to remove; instead, all the branches are cut off about eight inches above the cordons. Conditions permitting, this begins in mid-January, which leaves sufficient time for the hand pruning which follows to be completed by Spring. The process is simple: one worker walks down the rows with the hedge clipper, sawing off branches as he goes; then a second follows, yanking the just-severed branches away from the top wire, where they are still clinging by their tendrils. The branches are piled in the lanes between the rows, and are later collected and discarded. Anyone with the good luck to be at the vineyard while this work is in progress can claim all the cuttings he wants, free for the asking. He or she can take a bunch home, dry them out and use them to grill dinner, just as country-folk have done since time immemorial in La Mancha, Tuscany and Provence. The smoke imparts a delicious flavor to the food, and one might even accompany the meal with a wine made from grapes which grew on the very same vine.

The rest of the pruning is done by workers with hand clippers and long-handled lopping shears. Their objective is to leave a series of spurs spaced five to eight inches apart all along the upper surface of the cordons. Most often there will already be a branch in the spot where the pruner wants the spur to be, since these vines have been pruned in the same pattern dozens of times before. In that case, the worker simply cuts off the branch just above its third bud from the bottom and moves on to the next.

Frequently all three buds on the spur will sprout a branch in the spring. If none of those branches break off or get cut off during the summer, next year's pruner will find three branches growing in roughly the same spot and will have to choose which one to keep. Ideally, he will find one which emerged from the very bottom of the spur. It will be healthy-looking and growing straight up, rather than slanting downwards or to the side, and that's the one he'll save. The other two get cut off.

Grapevines have a natural tendency to concentrate their most vigorous growth at their far ends and to gradually stop producing branches further back on the cordon. When the winemaker sees this happening, he looks for a branch growing close to the vertical portion of the vine which can be converted into

a replacement cordon. Nature frequently obliges by providing a large and vigorous cane in the right spot, thicker in diameter than the other branches and
growing sideways, low enough for the power trimmer to have passed right over
it. The winemaker bends this into position alongside the faltering cordon and
attaches it to the wire with plastic ties. If it happens to have well-placed branches growing on it, all the better! He shortens them to the standard three-bud
spurs; if not, its buds will produce branches over the summer, and they will
be pruned the following year. Eventually, the original cordon will be severed
and discarded, and the new one will take up the slack without missing a beat.

Dead Vines and New Ones

Grapevines, like all living things have a finite lifespan, and although the
vines in this particular vineyard are as yet too young to die of old age, a certain
few succumb each year to mechanical or insect damage or to disease. During
winter they are visually indistinguishable from the dormant vines around them,
and thus go unnoticed until the worker pruning them cuts into only hard, life-
less and dry wood. Once a dead vine is identified it is cut off near the ground,
leaving an empty space in the row – a space which might otherwise produce
enough grapes to make a few bottles of wine. Obviously, the winemaker will
want to replace the vine, but for the time being there is little he can do but at-
tempt to stretch the ends of the adjacent vines into the gap.

The traditional method of creating new grapevines is to grow roots on cuttings taken from an existing vine – an elementry form of cloning which could
be easily accomplished at most vineyards. (It is also possible to grow vines from
seeds, but cloning is faster and exactly duplicates the grape which the vintner
wants, while a vine grown from seed rarely does.) For most of the centuries that
humans have cultivated grapes, vintners would cut branches in the fall from
vines of the varieties they wanted to propagate, tie them in small bundles and
bury them over the winter in sandy soil. When they dug up the cuttings the
next spring, a few would be found to have sprouted tiny roots. These would
be transferred to a nursery bed to grow for a year, and then be planted in the
vineyard the following spring.

Even today, vines produced in this manner are filling empty spaces in vineyard rows in many parts of the world; but when dealing with European wine

grapes in the eastern United States, things are a bit more complicated. We have a native insect – a species of louse called *phylloxera* – which eats the roots of some grapevines. Not only does it destroy root systems by eating them, but the wounds it leaves by its chewing provide entry points for bacteria and fungi to wreak further havoc, all of which kills the vine over the course of a couple of years.

European grapevines are especially vulnerable to this pest, which is the main reason why Thomas Jefferson and other American would-be winegrowers were so unsuccessful at growing European grapes here. Certain non-European grapevines have roots which are resistant to the insect, so our healthy clone must be attached, or *grafted*, to such a resistant root before it may be safely planted.

The process of grafting is difficult and is usually left to specialized professionals who sell grafted plants to vineyards at pretty high prices and only in minimum lots. Of course, the person who does the grafting needn't bother growing roots on the cutting; instead, he just takes it from the vine and grafts it directly to the phylloxera-resistant root. He does, however, have to propagate his rootstocks. For that, he follows the same steps just described for cloning new vines, since a rootstock, after all, is nothing but a grapevine, albeit one that is more useful for its roots than its fruit.

When Mike has enough empty spaces to fill, he orders replacement vines from such a professional and plants them in the spring.

SPRING

Springtime comes to Nassau Valley in fits and starts. Cold, damp blustery weather alternates unpredictably with shirt-sleeve-only days when the wine-maker and his assistants can finish pruning those last few rows which remain undone at the end of winter. Spring in a vineyard is not a time of sweet scents and showy blossoms, as it is everywhere else in the Delmarva countryside; but it is nonetheless the time for rebirth and new growth, and even for a resurrection of sorts.

The first sign that the vineyard is breaking its winter dormancy occurs when the ends of recently pruned spurs begin to exude a thin sugary liquid. This is water extracted from the ground by the awakening plant; it courses through the vine transporting the nutrients which will promote the swelling of the buds and the first surge of new growth. Those vines which were pruned after they broke their dormancy haven't yet formed dry callouses over their pruning cuts to block the sap from dripping, and so it flows now like water from a leaky faucet, dissipating the benefit the plant would otherwise obtain from that first surge. It is specifically to avoid this that vintners race to finish their pruning while winter is still upon them. Nevertheless, it is startling to see the quantity of the plant's lifeblood that can be spilled without doing any permanent damage to the vine. It is part of the miracle of the season that the cuts soon scab over despite the pressure of the pumping sap, and the plant quickly recovers from its loss.

As the season advances, the buds on the spurs – which had been barely large enough to notice during the winter – swell to nearly the size of kidney beans and then split open to reveal tiny green shoots. Each shoot comes with a pair of diminutive leaves, pale green tinged with pink, directly opposite each other and perfect miniatures of the leaves the vine will produce throughout the coming summer. The shoots quickly become branches and produce more and bigger leaves, as well as the wispy tendrils needed to grab onto trellis wires. And then at last, there appear on the branches the things for which the vineyard was created in the first place: tiny flowerlets shaped like miniscule clusters of grapes. These are called "inflorescences." By mid-to-late May these will open for pollination and over the summer will become the grape clusters from which wine is made.

Above: Inflorescence

Below: Same open for pollination

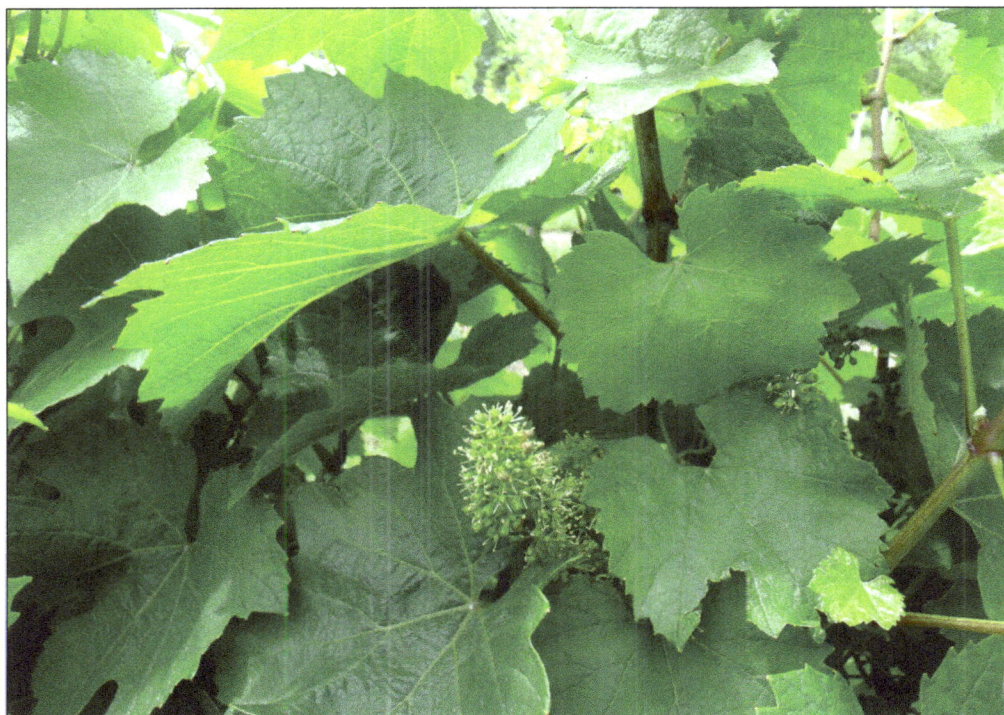

And now for one last miracle; remember the dead vines? Many of them – by no means all – will produce a bud somewhere down on the little stump which was left in the ground when the cordon was cut off. When a shoot emerges from the bud, the winemaker will tie it to a stake to keep it vertical, wrap a plastic shield around it, and nurture it over the coming summer and the summer after that to replace the vine out of which it grew.

Traditionally, spring is the favored season for planting a new vineyard, a back-breaking, multi-month project of shifting soil with shovel and plow, lugging around crates of grape-vines as well as coils of trellis wire and other building materials, scraping mud from boots and clothing and sleeping very soundly at the end of each laborious day. The process begins by turning over yard-wide swaths of earth to the depth of a foot or two for the length of each row, mixing topsoil and organic matter with the sub-soil and dealing with whatever happens to lie beneath the surface of the ground. (When the Raleys planted this vineyard several years ago, they came across the remnants of a nineteenth century glass and pottery works in the middle of what is now a section of Cabernet Sauvignon.) The next step is drilling holes in the ground with an auger and placing thousands of expensive little vines in them one at a time, spreading their roots just so and putting each into the ground at precisely the right depth. And then comes the final step: looking for ways to keep the new vines from drying out when the nearest water source is hundreds or, for some vines, thousands of feet away from the plants. Happily for Nassau Valley Vineyards, the days of doing all that are many years in the past, except, of course, for planting some replacement vines every few years. Now, although spring is still a time of momentous activity in the vineyard, very little of it requires human labor; nature does it all! That being the case, the action moves indoors from the vineyard to the wine cellar.

Racking, Blending and Bottling

Spring is one of two periods each year in which there is little to demand the wine-maker's effort and attention in the vineyard; the other is that bit of winter which falls between the end of one year's wine pressing and the beginning of the following year's pruning. Mike fills these blocks of time doing three of a wine-maker's essential tasks: racking, blending and bottling.

When grapes are crushed and pressed after each season's harvest, microscopic pieces of fruit, pits and skin remain suspended in the newly-made wine. The wine is stored in large stainless steel vats, and as the winter progresses gravity causes those solids to sink to the bottom of the vat. "Racking" is the process of transferring the clarified wine in the upper portion of the vat into another container – usually another large vat. In former times, this was done by siphoning; in modern wineries an electric pump and accordion filter are the tools of choice. Once the wine is transferred, the sludge on the bottom of the vat is cleaned out, the vat is washed and steam-sterilized and becomes the receiving vat for the racking of some other wine. Most new wines are ready for racking by mid to late December of the year in which they were fermented; virtually all are ready by the time spring rolls around, although it is not especially unusual for a wine to require more than one racking before it becomes clear enough to bottle.

Blending, quite obviously, is the mixing together of two or more different wines in the hope of producing a drink that is better than its separate components. The key tools for this operation are the winemaker's palate and his good judgment. He already knows from past experience which wines blend best with each other, but their optimum proportions are not necessarily the same from year to year. To get those right, he has to assess the color, aroma and taste of each, and to do this accurately the wines under consideration must be at least moderately clear. That is why blending is usually done in conjunction with racking. Once the wines are blended they are returned to vats or transferred to oak barrels to age together in either place until they are ready for bottling.

As a general proposition, the winery's most distinguished red wines are not bottled for at least two years after their first racking, while most whites can be bottled as soon as they are clear. Within those parameters, the determination of when and whether a wine is ready to be bottled is one of the most important decisions that falls to a winemaker. Once he decides that the time is right, that day becomes a very busy one at the winery.

First, an elaborate network of hoses and a pump is set up to connect the vat containing the wine to the six-spigot bottle-filling machine. Then cases of sterilized bottles are stockpiled on one side of the machine, and empty cases lined up on a work table on the other side to receive the bottles as they are filled. When the pump is turned on, the person doing the bottling takes empty bottles

from the cases and places them on the machine's spigots with a slight upward pressure which starts the wine flowing. As each bottle is filled, the flow from that spigot turns off automatically; the bottler removes it from the spigot and puts it into an adjacent machine which instantly inserts a cork. She places the corked bottle into a case on the work table, while reaching simultaneously with her other hand for an empty bottle to place on the spigot which was just vacated. As she goes sequentially along the spigots, dexterously moving bottles from case to bottling machine to corking machine to other case, helpers are continuously putting new cases of empties on one side of the machine and carrying away filled cases from the other. They stack the filled cases five deep on pallets, stamp them with the name of the wine and wheel them to a section of the work floor away from the action. Labeling of the individual bottles will wait for another day.

SUMMER

The vines make rapid growth in the heat and humidity of the Delmarva summer. The tiny inflorescences visible in the spring have become recognizable clusters of hard little green grapes, and the shoots which earlier emerged from the buds on the spurs are now six or seven feet long and growing rampantly. Pruning the vines is almost as important during this season as it was in the winter.

"Suckers" – new plants growing off the roots of the vines – must be pulled or cut out before they start to steal water and nutrients from the established vines. Overly long canes and a superabundance of leaves can act like a tent, smothering the grapes in a damp blanket ideal for the growth of fungus and plant disease. These too must be trimmed back to expose the grapes to the sunlight and fresh breezes they need to ripen properly.

The European grape varieties planted in this vineyard are obviously not

native to coastal Delaware, and require special cultivation practices to protect them from the competition of our local insects, weeds and diseases. Carefully controlled spraying with agricultural chemicals is necessarily part of this regimen, although winemaker and vineyard manager, Mike Reese limits his spraying program to the unavoidable minimum, seeking to use alternative methods wherever possible. A good illustration of his attitude involves Japanese Beetles. These voracious eaters emerge from the soil in late June and immediately start eating the tender new leaves at the ends of the youngest branches. Because they have no natural predators in this region, the only way to stop them is to spray pesticides; but instead of doing that, Mike lets them gorge themselves on these too-long branches which he intends to cut off anyway. In late July the beetles reach the point in their life cycle when they stop eating and burrow into the ground. Mike then removes the badly chewed foliage they leave behind. The grapes are none the worse for the experience, and the whole problem is resolved without the use of any pesticides.

As the winery goes through the summer, it also becomes the venue for many extra-curricular activities – weddings, concerts, art shows, and a Sunday Farmers' Market. The weddings started coming to the vineyard back in its early

Trimming overly long canes

days when the enterprise was still touch and go. From out of left field, couples who liked the looks of the place started asking about having their weddings on the premises. At first, makeshift accommodations were the rule, such as rolling barrels out of the way to make room for tables in the wine-cellar. Later, two event centers were erected, built of custom-made bricks in the style of the 18th century carriage houses in Colonial Williamsburg – a favorite vacation spot for Peggy's parents.

✣ ✣ ✣

Making wine is fundamentally a matter of converting the sugar in grapes into alcohol. The percentage of sugar in a grape starts the season at zero and increases as the fruit matures to whatever is the highest percentage nature permits – usually somewhere in the low to mid 20's – at which point the grape is said to be "ripe." Experience gives the winemaker a fairly good idea of when ripening will occur, but he needs to know as exactly as possible in order to orchestrate the most elaborate event of the year: the harvest. So around the middle of August, he starts measuring sugar content with a hand-held refractometer. He selects an average-looking grape, crushes it on the lens of the device, and reads the sugar level from a scale imprinted on the lens. Once this tells him that sugar content is within the ballpark, he takes larger samples and analyzes them in his lab, hoping to predict the date of perfect ripeness, which will also become the date of the harvest.

Meanwhile, also in mid-August, a new and urgent issue confronts the vineyard – BIRDS, the mortal enemy of all winemakers. Now that the not-quite-ripe grapes are plump and translucent, birds of every feather converge on the vineyard hoping for a feast. They can destroy an entire crop in just a few days, unless the winemaker finds a way to head them off. This he does by installing a solar-powered loudspeaker which blasts out the screeches of predatory hawks at appropriate intervals. The device scares the birds away for perhaps a week, but keeping them away is not quite so simple. They soon start raiding around the edges of the vineyard farthest from the hawk sounds. If not checked at this point they will work their way inward and devour most of the crop, so the winemaker counters by adding another screech machine and a couple of cannon-boom machines to the threatened areas. Later he follows that by stretching reflective silver tape directly over each row in the vineyard to make wierd humming sounds while shimmering and flashing in every breeze. He can't install all the gadgets at once, as most birds quickly lose their fear of startling things which don't actually hurt them. So the winemaker must introduce new devices incrementally, hoping to harvest his crop before the birds get used to them. Who knew that out-smarting creatures with bird brains could present such a challenge?

There is a final group of vineyard chores which precedes every harvest. The first in the sequence is "de-leafing" – literally ripping leaves off the vines a week or two ahead of harvest. The purpose of it is to expose the grape clusters to the air and sunlight which they need to ripen uniformly and completely. This seemingly mindless task actually demands considerable judgment by the worker performing it. The same leaves which block the sun and inhibit air-flow also photo-synthesize the very sugar wanted in the grapes. They are, in Peggy's words, "the plant's solar panels", so the amount and condition of the foliage must be carefully assessed to avoid removing too much. One thing which makes

Weapons in the battle with the birds

the assessment a bit easier in this particular vineyard, is the fact that the long rows of vines run roughly north-south. The grape clusters on the east face of the vines benefit mightily from leaf removal, which enhances their exposure to the gentle morning sun, while the opposite is true for those grapes on the west face, where the leaves protect them when the sun becomes scorching in the afternoon. In the photo below, which was taken in mid-afternoon with the camera looking northward, we can clearly see that the east face of the vines at the left of the picture is nicely shaded, while the west face of the row on the right is fully exposed. Naturally, many more leaves are pulled from the east than from the west faces of the vines.

De-leafing

AUTUMN

Technically it is still summer when the first grapes ripen at the end of August, but in the vineyard, the transition to autumn is signaled, not by the equinox, which is still three weeks away, but by the harvest. When we reach Labor Day, that harvest is close at hand. It is a time of both anticipation and anxiety at the winery, the time when all the work of the past year should yield its reward, and also the time when a flock of migrating birds, a hurricane or even a big rainstorm – none of which are unusual at this time of year – could ruin everything.

The vineyard staff goes about the first days of September working on those tasks which must be done immediately before the harvest. The crusher and wine-press are taken out of storage, given a cleaning and set up on the wine-cellar work floor, while the call goes out for extra workers to pick the grapes later in the week. Outdoors, in the vineyard, the grass between the rows

Chardonnay

gets a short mowing, and workers go through the *Chardonnay* section – scheduled to be the first picked – removing imperfect grapes from the clusters. Winemaker Mike Reese and vineyard co-owner Peggy Raley-Ward devote themselves to everyday tasks while stoically aware of the truth which nature teaches all farmers – that there is little they can do to prevent the several possible disasters which might strike at this vulnerable moment and bring to naught a year's efforts and investments.

This year's (2016's) first disaster was birds in the Chardonnay. During their mid-August battle of wits with the human team, they managed to consume all but twelve-hundred pounds of Chardonnay grapes from a vineyard section which often produces three or four tons. On harvest day, the pickers filled their picking lugs with the shockingly meager crop which remained and loaded them onto a trailer for the short trip from vineyard to wine cellar.

Determined to prevent any similar crop loss in the future, Mike resolves to cover the Chardonnay with nets next summer. This means draping or wrapping each row of vines with plastic bird nets, hundreds of feet long. The nets

Merlot

are expensive, awkward to install, and frequently get snagged on one piece or another of vineyard machinery. Leaves and branches grow through them and tendrils grip them while they are on the vines, making them difficult to remove when the time comes to actually pick the grapes. Nevertheless, they are the only real barrier one can erect between the birds and the fruit. They are not a perfect barrier. It is probably impossible to completely seal every seam, and every so often some daredevil bird will smash through the net; but the measures taken this year obviously didn't work, and our winemaker is not a man to twice come up shy for want of trying something different.

In a year, we'll know if nets are the answer. For now, the Chardonnay grapes are kept overnight in the wine-cellar's cold storage room after being transported from the field, and the next morning the wine making begins. Step one is pouring the grapes into the crusher, a machine which doesn't so much crush the grapes, as break them open and detach them from their stems. The stems are forced out a chute and into a waiting container, while the mangled grapes and their juice fall through the bottom of the machine into a twenty-gallon receptacle. As soon as three receptacles are filled, the crushing is interrupted long enough to empty them into the basket of the wine press.

Pressing is the second step, and is done to squeeze as much juice as possible from the crushed grapes, while straining it to remove pits and skin. The strained juice is continuously pumped into the container where the actual fermentation will take place, usually a 750 gallon fermenting vat. But today's volume is so small, it goes instead into fifty-five gallon drums in the cold storage room. While the pressing is in progress, the winemaker and his assistants begin crushing the next batch of grapes.

(It is important to note that what is being made here is white wine, so the pressing proceeds immediately after the grapes are crushed, and only the juice is fermented. When making red wine, the crushed grapes are not pressed until around a week later.)

After each load of grapes is pressed, a layer of compressed skins and pits – known as the "cake" – remains in the basket . This is scooped out and discarded, and the press is then ready to receive its next three receptacles of crushed grapes.

This two-step process is repeated until all the juice is in the two drums which are all that is needed to contain it. Twelve hundred pounds of grapes yield only eighty-four gallons of juice. Equally frustrating, Mike's analysis of the juice shows it to be of extraordinary quality. This will be a banner year for Chardonnay.

The press basket is disassembled for cleaning, the crusher, press and containers are sterilized and the work-floor washed. The big piles of cake and stems are added to the soil in a distant corner of the property, while the drums remain overnight in the cold storage room, which is deliberately kept too cool for fermentation to begin.

The following day, the juice receives its first racking. Although yesterday's pressing screened out most of the solids, nearly half a gallon of fine particles remained suspended in the juice. During the night most of those sank to the bottoms of the drums. These are eliminated by racking and the racked juice is now close to completely clear. Mike again leaves the drums overnight, this time in the slightly less cool main workroom where the juice will warm to the temperature he considers ideal for fermenting, and the next morning he will inoculate the juice with yeast to start the fermentation.

Fermentation is a chemical process which converts sugar into alcohol. It requires oxygen and warmth, and the more of each you have, the faster and hotter the fermentation will be. For the flavor and texture which Mike envisions for this Chardonnay, a slow and relatively cool fermentation is required. He stirs yeast into the juice and closes the drums, leaving a few inches of head space at their tops to provide some oxygen. He then fits each drum with an air-lock to prevent pressure build-up in the drum by allowing air to escape. Carbon dioxide is a by-product of fermentation, and as it is produced, it displaces the oxygen in the drum and pushes it out through the air-lock. If all the oxygen were to be expelled in this manner, fermentation would cease prematurely; Mike averts this by opening and re-closing the drums occasionally over the next few days, which reintroduces all the oxygen needed to keep the fermentation going at the desired rate.

Now attention shifts to the next grape in the ripening sequence, the *Merlot*. A few days before the harvest, workers go along the rows of Merlot removing and discarding individual grapes from the bunches. Two categories of grapes are removed: green ones, which are unripe and therefore will add too much tartaric acid and not enough sugar to the mix; and those with pierced or broken skins, which can add unwanted organisms and odd tastes to the wine. Meanwhile, Mike analyzes samples and schedules the harvest for the coming Wednesday.

Just before sunrise on the appointed mid-September Wednesday, a dozen pickers assemble at the winery. They are a mix of regular vineyard employees, their friends and relatives, and residents of the area who think harvesting in a vineyard might be a fun way to spend a morning. (And indeed, it is!) Everyone in the group will be paid the standard local wage for part-time agricultural labor and given lunch in the vineyard. Mike gives the group some brief instructions, provides each picker with a pair of clippers and sends them off to work. Each picker selects a row of grapes, and works his or her way from one end of it to the other, cutting clusters of grapes from the vine and tossing them into the yellow plastic baskets, called "lugs," which had been set out earlier in the rows. You push your lug along as you go, and when – after picking three or four vines – you have filled it, you find an empty one strategically placed just where you need it to continue working without interruption. It takes about four hours on this beautiful sunny day to pick the entire crop of Merlot – just shy of one ton.

The vineyard tractor pulls a low trailer through the rows to collect the lugs and haul them to the wine-cellar, where the crusher is already in place for the next operation.

A bunch of pizzas and several six-packs of ice-cold Gatorade have been set up on a large picnic table near the wine cellar, and all enjoy a convivial al fresco lunch. Then the group breaks up, with the regular winery workers assembling in the cellar to crush, and everyone else returning to their normal day's occupations.

Unlike the Chardonnay, the Merlot – as well as all the other red-wine grapes – will be fermented in open tubs called primary fermenters. These are fiberglass cubes, enclosed in tubular metal baskets to keep them stiff. Each cube holds a ton of grapes. The crusher is mounted on top of the primary fermenter, which is seated on a wooden loading pallet to facilitate moving it. This arrangement of equipment places the crusher nearly six feet off the floor, so a platform is rigged alongside for a worker to stand on while he dumps the grapes from their lugs into the crusher. In olden times, crushing grapes was a communal effort, with barefoot villagers clambering into wooden fermenters to tread on the grapes, much like Lucille Ball did in the classic *I Love Lucy* episode. Nowadays the job is done by electric crushing machines. That makes the process more sanitary, but nowhere near as much fun.

In any event, crushing is the first step in making any kind of wine, and the Merlot grapes are duly crushed. Because this is to be a red wine, the next step is fermenting the grapes "on their skins", in other words, without separating the juice from the rest of the crushed grape. Grape juice is a nearly colorless liquid. The pigments which make a wine red are all concentrated in the skins, and are absorbed into the juice during this primary fermentation. Other compounds in the skin, pulp and even the pits give the wine its flavor, and help to brighten and clarify the finished wine.

If the crushed grapes, now referred to by the term "*must,*" are simply left alone for a week or so at this point and then pressed, the end result will be something which may legitimately be called wine. That, in fact, was pretty much the method used by the ancients. Over the past few centuries, however, winemakers have found ways of introducing sulfur dioxide into the must to disinfect it; so Mike mixes in the precise amount of SO_2 needed to purify the volume crushed without affecting the flavor of the finished wine or harming the yeast he will introduce in a few hours to start the fermentation.

Next, he checks the levels of sugar and acid in the must. The amount of sugar determines the alcoholic content of the wine. Mike is looking for a sugar level of 25 to 26 "*brix*," the scale on which grape sugar is traditionally measured. The brix number is approximately double the projected alcohol content of the finished wine, so a brix of 25 or 26 will yield 12 ½ to 13% alcohol. In this case, the brix is just right. Had it been a little low, the problem could have been easily corrected by adding a measured amount of ordinary sugar.

Tartaric acid is another organic compound which occurs naturally in all grapes. Mike measures it also, as it is an important element in the taste of the finished wine: too much, and the wine will be astringent; not enough, and the wine will be insipid and unstable. Here, the amount of acid is also right, which is not surprising. As grapes ripen, their sugar content increases while the amount of acid in them remains static. Typically, the two wind up in perfect balance when the grapes are fully ripe. Correcting an imbalance can be quite tricky, but is seldom necessary except when circumstances require harvesting grapes which are seriously under-ripe or over-ripe.

While the winemaker does this testing, workers pick through the must by hand to remove any little pieces of stems or leaves which might have snuck past the crusher screen, as well as any green berries which eluded them during the pre-harvest culling. Mike favors a cool fermentation for most of his wines, and has placed a floating thermometer in the fermenter. He checks it now and finds the temperature of the must is warmer than he wants, so he uses a pallet jack – sort of a cross between a fork-lift and a hand truck– to wheel the fermenter into a small room where he can raise or lower the temperature as needed. The must spends the night cooling off.

The following day, with the temperature corrected and the winemaker satisfied that the must will produce the wine he wants, he adds the yeast "starter" to the must. Yeasts are the micro-organisms which convert sugar into alcohol, and thereby convert grape juice into wine. They are designed by nature to float around in the open air and accumulate on the outer surfaces of grapes and several other fruits. There are many strains of yeast, and over time, vintners have selected and cultivated certain ones which they believe make the best wines. The earlier dose of SO_2 has stunned whatever wild yeasts were on the crushed grapes; they wouldn't do as good a job as the cultivated ones. By the time they recover, the better yeast will have capitalized on their head start to become the dominant population. To make the starter, Mike dissolves a pre-measured quantity of selected yeast in a quart or two of grape juice scooped from the fermenter, sets it in a warm, airy place for an hour or so and watches it bubble and churn. Then he pours it into the primary fermenter and puts our microscopic friends to work.

Active fermentation in an open container can take as little as a few hours or as much as a few days to get started – depending on the temperature of the must – but once it does, it is quite a thing to behold. Pretty pink bubbles float and burst on the surface of the must and the wine-cellar fills with the distinctive sweet aroma of fruit, yeast and alcohol mixed together, as well as quite a few fruit flies. The juice of the grapes sinks down in the fermenter, separating from the lighter skins and pulp, which float to the surface, where they dry out and stick together in a stiff crust which winemakers call the "cap." The cap can exclude the oxygen needed for fermentation, so the winemaker punches it down into the fermenter every few hours with a long-handled implement designed for that purpose. This aerates the must and keeps the solids uniformly distributed in the juice.

Mike keeps careful tabs on the progress of the fermentation. That

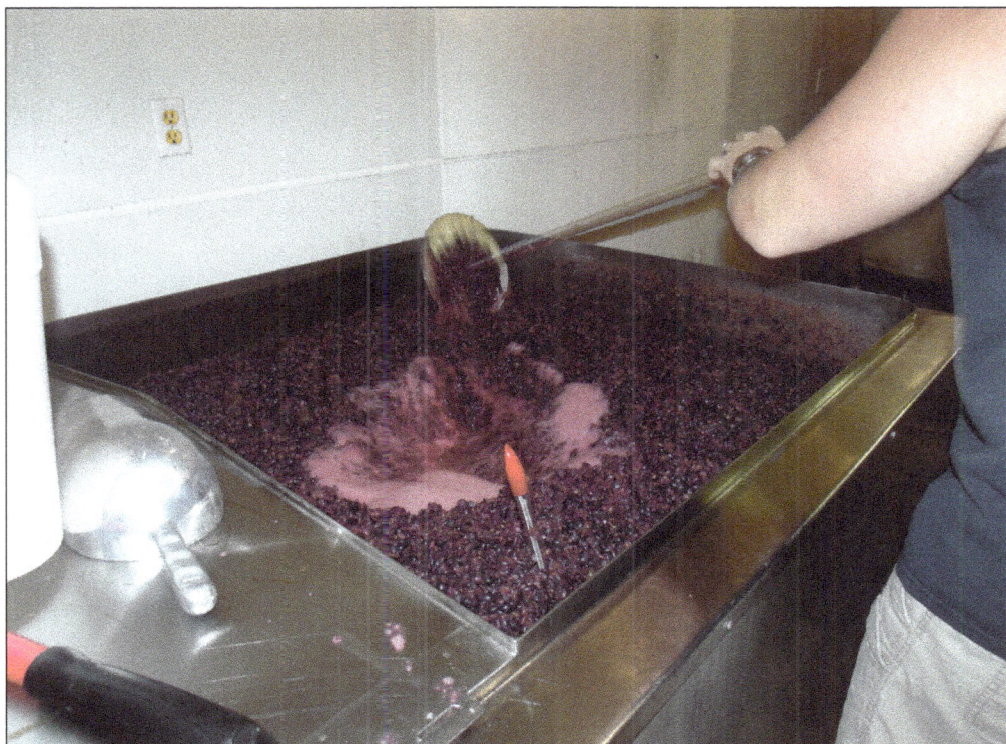

Above: Punching down the cap

Below: Checking specific gravity

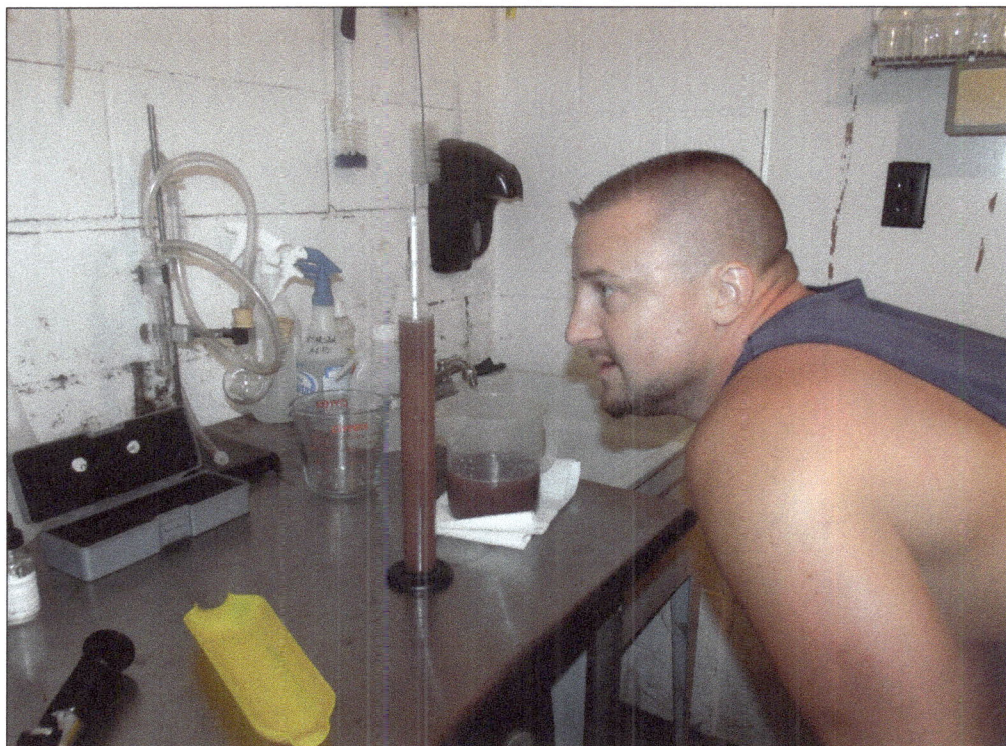

thermometer is still floating in the must, and the fermenter will make a few more visits to the temperature-controlled room before the process is over. When the must has fermented for a few days, Mike begins checking its specific gravity with a hydrometer. Specific gravity decreases as the sugar in the must is replaced by alcohol. To create a dry red wine, a winemaker has to eliminate all sugar from the must. After about a week, the bubbling stops and the hydrometer readings show a specific gravity consistent with the absence of sugar. This means that the fermentation is complete and it is time to press.

Pressing a fermented must is quite different from pressing fresh grapes, although the same press is used for both. Most of the juice originally in the grapes seeped out during fermentation, so the mixture in the fermenter is in large part pure liquid. The first step in pressing this doesn't involve the press at all; instead, a hose is lowered into the fermenter and the new wine is pumped into a closed vat for storage. To avoid clogging the pump with pits and other solids, Mike attaches a strainer to the end of the hose before inserting it into the fermenter. It is a device of his own design, which he made by drilling small holes in a length of PVC pipe. Once the pure liquid – generally known as "free run" – is out of the fermenter, the solids which remain are scooped into the press basket. Because these solids are totally saturated, copious amounts of wine

flow out of the basket without any pressure being applied. This too is free run. Finally, the press is turned on to squeeze the last of the wine out of the solids. Often, there will be a noticeable difference in quality between the free run and the pressed wine; when that occurs, the two will be pumped into separate vats after they go through the press. If the winemaker does not perceive sufficient difference between the two to justify that extra step and the additional storage space it entails, both will go directly to the same vat, to rest until ready for racking.

While all this is going on, Mike schedules the harvesting of the grape which will ripen next, the *Cabernet Franc*. This grape is a close relative of the noble *Cabernet Sauvignon*, though a little fruitier and with less tannic acid than its majestic cousin. After that will come the *Petit Verdot*, also a relative of the Cabernet clan, darker colored and, as its name suggests, smaller. Its tannin content is high because its smaller size creates a larger ratio of skin to juice in its berries. In France, both these varieties are generally blended with Merlot and Cabernet Sauvignon, to produce the perfectly balanced red wines of Bordeaux. Mike does the same thing here to make the winery's award-winning, *Indian River Red*. Last of all, around the beginning of October, the Cabernet Sauvignon itself will be harvested – the pride of the vineyard.

These final three varieties are managed much the same as the Merlot: first the unripe and damaged grapes are culled in the field, then the rest are harvested, crushed and fermented. The slightly different wrinkle with these later-ripening varieties is that there are fewer unripe grapes to remove this late in the season, but the incidence of damaged grapes is markedly higher. This is primarily due to insect damage, mostly from wasps. When the insect pierces a grape's skin, air enters providing the opportunity for whatever yeasts may be on the grape to start fermentation. This is bad because the ultimate end-product of fermentation is not wine, but vinegar, which is why fermentation at any winery is so carefully monitored. Outdoors, fermentation proceeds very quickly in the warm weather of early fall, and within a few days, many a grape will have turned sour.

The culling is messier and slower than with the earlier grapes, and the many wasps lurking in the grape clusters can be unnerving, although surprisingly, they pose little or no hazard to the bare- handed culler. With a good meal under their belts, and being a little tipsy besides, wasps are far less belligerent than usual. They won't sting unless you actually squeeze them.

This year, all went as planned with the later-ripening varieties until it came to the Cabernet Sauvignon, which were on the very cusp of ripeness when the Cabernet Franc and Petit Verdot were crushed. Mike had scheduled the harvesting of the Sauvignon to begin as soon as he finished pressing the other two. Then came the sudden forecast of severe rainstorms on the way. All available manpower was diverted to harvesting as much as possible of the ripest Cabernet Sauvignon, which turned out to be a total of five rows. This was barely completed when the downpour commenced, and it lasted five days, dropping more water on the Delmarva Peninsula than anyone could remember in such a concentrated period of time.

A rain delay at harvest time presents two specific problems. First, the vines quickly draw up the excess water in the ground and pump it immediately into the fruit, diluting both their sugar and acid, which ultimately leads to less robust flavor in the finished wine. The winemaker can partially compensate for the dilution by adding sugar and tartaric acid to the must during fermentation, but the other flavor components in the grape will still be diminished. The second problem is more serious: rot in the fruit. The wetness and dim light associated with prolonged rain present the ideal environment for disease to take hold and spread. When the rest of the Cabernet Sauvignon was finally harvested during a break in the weather, much of it was found to be unusable, and the remainder quite inferior to the Cabernet Sauvignon of previous years. Mike quickly made the difficult decision which the circumstances demanded: the vineyard would not bottle a Cabernet Sauvignon from this vintage. Ordinarily his top of the line grape, the Cabernet Sauvignon would this year be just another component of the vineyard's *"Redneck Rouge,"* the every-day wine at the bottom of the price scale.

Mike calculated the amount of sugar and tartaric acid needed to make up for all the dilution and added it to the must, which, perversely enough, was so voluminous that it had to be fermented in the winery's biggest fermenter, and stored thereafter in one of the cellar's largest tanks. To store wine in bulk after fermentation, the first step is to pump carbon dioxide into the storage tank. Being heavier than air, the carbon dioxide forces the air inside the tank upwards. When wine is subsequently pumped into the tank it displaces the carbon dioxide, pushing the gas upwards and compressing it, which, in turn, further compresses whatever room air remains in the tank. When the pressure reaches a certain level, Mike opens a relief valve to allow air to escape. Once the process is complete, the tank is nearly filled with wine, and whatever head-space remains is occupied by the antiseptic CO_2.

By late October, the last grape has been pressed, the equipment cleaned up and put away for the year, and the wines are in storage vats. The different varieties – even those destined to be blended – remain in their individual vessels to mature separately. While there, "resting on their lees", they will undergo some chemical changes in a process called malolactic fermentation, which is essentially the conversion of the malic acid found in grapes and many other fruits into lactic acid. This mellows the taste of the wine, and is initiated by introducing a yeast specific to that purpose. The winemaker will look in on the new wine periodically to make sure everything is all right and to inform the blending choices he will make in the next few months.

AUTUMN PART 2, The Non-traditionals

By early November, the work in the vineyard itself is finished for the calendar year, but there is still quite a lot of wine to be made – from fruit not grown on the premises. Each year the winery buys hundreds of gallons of grape juice from growers in the Finger Lakes region of New York. The grapes, a native American variety, some hybrids and a Pinot Grigio are crushed and pressed where they are grown, and their juices shipped here in refrigerated trucks. The common characteristic of all these different grapes is that they grow best in more northern locations than Delaware, places where the air is cooler and drier, and the higher latitude provides more hours of daylight during the summer. Mike makes several wines from these juices, following the same steps described earlier for wines grown in the vineyard, but with the difference, of course, that the process does not involve crushing and pressing.

The winery also makes blueberry and peach wines. These are also made from juices purchased from growers of those fruits and shipped here as liquids. Making wines from fruits other than grapes presents a few special challenges, but is not essentially different from making grape wine.

Several varieties of grapes are native to North America. All of them are as different from the European varieties as lemons are from oranges. That's because, like lemons and oranges, they are actually different species, although they have many similar features and are members of the same genus. Wine derived from North American grapes generally has a distinctive spicy sweetness about it, which people in the wine business call "foxiness". This is usually quite pronounced, often unpleasant, and, in any event, very different from the taste one expects to get from a glass of wine. There is, however, one native grape in which these characteristics are barely noticeable, and in fact actually enhance the flavor of the wine. That is the *"Delaware"* grape; it makes a sweet white dessert wine, which is a perennial crowd pleaser at shows and wine tastings. Although Delaware would be proud to claim this grape as a native of the state, it actually gets its name from a town named Delaware in central Ohio, the place where it was first discovered. It is the only native grape used at Nassau Valley.

A second notable difference between American and European grapes is the one which brought about the creation of the *"French Hybrids."* There is an American grape species named *Vitis Labrusca*, which is native to our northeastern states. Its roots are stringier and far tougher than those of the species *Vitis Vinifera*, to which all European grapes belong. During the nineteenth century, some plants of the labrusca species – most likely *Concord* grapes – were exported to France and carried with them the same villainous bug mentioned earlier in these pages, the root-chewing phylloxera. This minor pest for the tough-rooted labrusca vines proved to be a devastating scourge to the vinifera. It wiped out the French wine industry for over a generation, and with it the livelihood of a good part of the country. Scrambling to find a cure, vintners tried grafting vinifera branches onto labrusca roots, but the two species were not wholly compatible and the grafts didn't take. So then scientists sought to create root-stocks through hybridization between the two species. They attempted thousands of crosses and finally achieved a few successes – vines which would accept vinifera grafts and which also had roots tough enough to withstand the phylloxera. They saved the vineyards of France, and coincidentally made it possible for Nassau Valley Vineyards to grow vinifera grapes in Delaware. Every vine in this vineyard is grafted on a hybrid rootstock of one sort or another.

Although those 19th century scientists were interested in roots rather than fruit, some of their hybrids turned out to produce grapes which closely resembled their vinifera parents, so vintners in France began cultivating them specifically for wine. Unfortunately for them, French consumers showed no more enthusiasm for these wines than Americans did for the Edsel and the new Coke in the 20th century. In a land where wine is so ingrained in the culture, few could see any point in drinking the hybrids once the originals were back in production. So hybrid grapes were relegated to use primarily as bulk fillers to increase the volume of ordinary wine made from traditional grapes. That might well have been their permanent fate, were it not for the intervention of the late Phillip M. Wagner, a noted authority on growing grapes and making wine, and a prolific writer on the subject.

Wagner was the European correspondent for *The Baltimore Sun* in 1936 and '37, and became familiar with the hybrids while on that assignment. Recognizing that roots which could withstand phylloxera in France should logically do the same in America, he imported several varieties of hybrid grapes for his vineyard in Maryland. He found that many of them made excellent wine and, most importantly, could be grown on their own roots with no need for grafting. He began propagating many of the best varieties on a large scale and marketing them to home and commercial vintners all over the northeast. Among these were *S-V 5276* (the five thousand two hundred seventy-sixth cross attempted by French horticulturalists Bertille Seyve and Victor Villard) and *Vidal 256*. These two spectacular grapes – now with the hip names, "*Seyval*" and "*Vidal Blanc*" – can be blended together to make a delicious dessert wine. Peggy developed that combination early in the history of her winery with some assistance from Phillip M. Wagner. It is now one of Nassau Valley Vineyard's proprietary blends and is used to produce the vineyard's *Meadows Edge*.

Another French Hybrid whose juice the vineyard purchases is the *Chambourcin*. It is a red grape, but coming as it does, in the form of juice, Mike ferments it in a closed fermenter, just as he does his white wines. This yields a lighter colored red wine than would be the case were it fermented on its skins. How is the wine red at all, if it isn't fermented on its skins? The growers heat the grapes slightly at the time of crushing, which quickens the absorption of pigments by the juice. Chambourcin is mixed with *Cayuga*, a white hybrid devel-

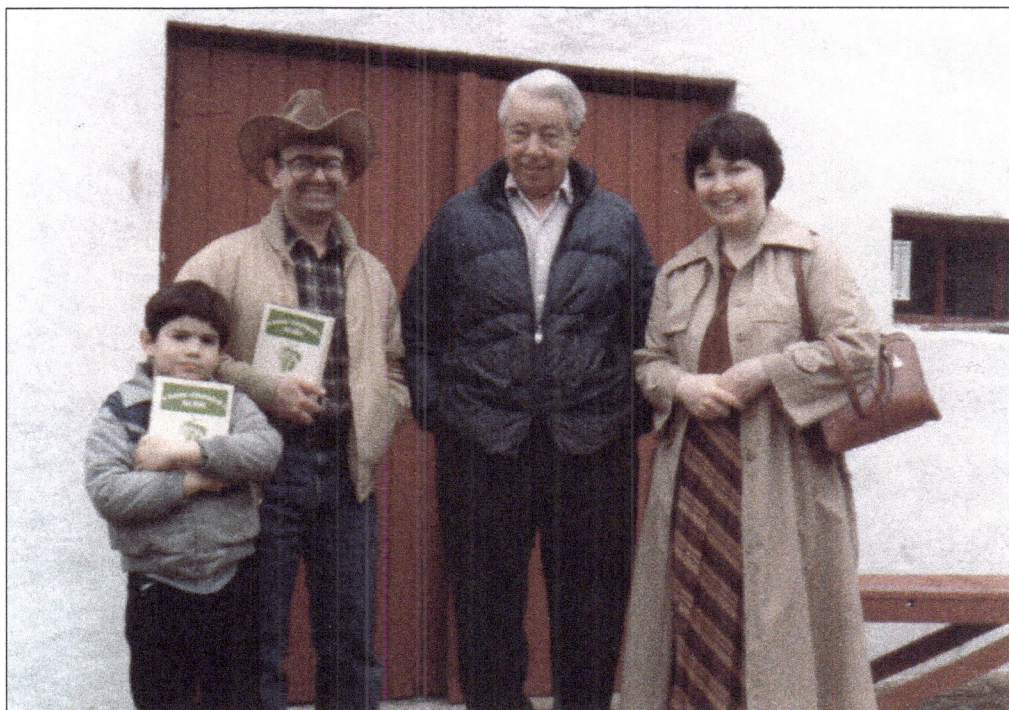

Phillip M. Wagner in 1984 with the author and the author's wife and son.

oped by the New York State Agricultural Experiment Station in Geneva, New York. This produces a fruity and slightly sweet pink wine - a traditional French style rosé. Nassau Valley's name for the combination is *Cape Rosé*.

Finally there is the *Pinot Grigio*, a vinifera familiar to us all, and possibly the best selling wine in the English-speaking world. The grape gets its name from the greyish tints in its purple skin, and is one of the few red grapes which vintners press right after crushing in order to make a white wine.

And so we end this year in the vineyard, as well as this little book, pretty much where it began, with all the grapes harvested, leaves fallen from the vines and long chilly nights pushing aside the shortening days. A better ending to the story might show us popping the corks of the wines which were made in the preceding pages, but that is an event for another year. For now this vintage must wait in the cellar as time – the final ingredient of wine-making – turns them into something worth drinking. Any year in a vineyard is part of a repetitive cycle which should always end at the same point chosen for its beginning. Future years may give us wines which are better, worse or the same as those of 2016, but the ancient art of making them will go on just as it has in these pages.

EPILOGUE 2017

So the vineyard bought the nets – enough to cover about ninety percent of the Chardonnay. In their first season, at least, they presented none of the problems I described earlier. Over four tons of Chardonnay were harvested in 2017, from which more than 500 gallons of wine were produced. The ten per-cent of the Chardonnay vines which were not covered by nets – a control group of sorts – also yielded grapes which are included in the four ton total; however these vines suffered losses from birds comparable to those of the previous year.

www.ingramcontent.com/pod-product-compliance
Lightning Source LLC
Chambersburg PA
CBHW040143200326
41519CB00032B/7589